Glass is like a riddle. It is hard but it can break. It is made from sand, but you can see through it.

We can find glass all around us. Windows are made from glass. Mirrors are made from glass.

Believe it or not, glass is made from liquid sand. You can make glass by heating up sand to a very high temperature until it melts.

When molten sand cools, it changes into glass. When sand is molten, you can shape it.

You can pour it onto a flat surface to make sheets of glass.

Glass is all around in our homes. Bottles, jars, mirrors, watches and windows.

Glass is one magic material that we use all around us.